Contents

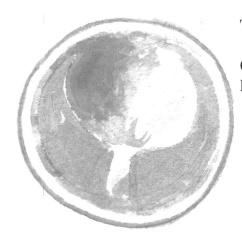

SOAP SMARTS

"Are you washing your face?" your parents call.
"Yes, I'm washing my face!" you tell them.
"Are you using soap?"
"Um ..."

Why do you have to use soap when you wash up? How does soap clean your body, your clothes, or your family's car? What makes soap and water so powerful? This book shows you the science of how soap works, plus a whole lot more.

Soap is neat, as well as clean. Just add water, and soap becomes suds, slime, and (for the most fun of all) bubbles. Soap can be dry or slippery, liquid or solid. And you can use soap to understand electricity, light, and many other science topics.

If you find a soap science word you don't understand, look it up in the Glossary on page 63.

4

Mix oil and water

During your shower you drop the soap. Can you clean yourself just as well with plain water? Dry off and do this experiment to see how soap works.

You'll need:

- 2 small jars with tight lids
- water
- red food coloring
- cooking oil
- liquid soap or liquid dishwashing detergent

1 Fill one jar with water. Stir in two drops of red food coloring. Pour half of the pink water into the other jar.

2 Fill the rest of each jar with cooking oil. Watch how the liquids separate into two layers, with the oil on top.

3 Add a squirt of soap to one jar. Screw both lids on tight.

4 Shake both jars for the same amount of time, and set them down again. How do the two liquids look now?

WHAT HAPPENS?

When you shake the jars, the liquids are thrown together like football teams at the line of scrimmage. In the jar without soap, the oil and water eventually settle back into their yellow and pink layers, like teams going back into their huddles. In the other jar, however, soap doesn't let the oil and water separate, so you see a sudsy orange mixture.

Soap, water, and oil (and everything else) are made up of tiny particles called molecules. Oil and water molecules slide off each other, but soap molecules stick to them both. When the molecules mix, one end of a soap molecule grabs water molecules, and the other end grabs oil. That makes it difficult for the liquids to separate.

Soap works the same way on your body during a shower. When something oily, such as french fry grease, is stuck to your skin, plain water just slides off it. But soap holds on to the oil and then is carried by the water down the drain.

What happens if you pour oil on soapy water? What if you pour water on soapy oil?

How can you test whether the food coloring changes the results of this experiment?

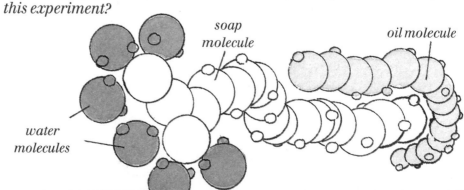

soap molecule

oil molecule

water molecules

Ocean bath

A dip in the ocean is more fun than a shower! How come you can't save time by taking a bar of soap into the ocean and bathing in your bathing suit?

You'll need:

- sea water (if you don't have any, see step 1)
- a paper coffee filter
- a filter holder
- 2 glasses
- tap water
- salt (see step 1, if necessary)
- powdered soap or powdered detergent
- 2 spoons
- 2 straws
- 2 mouths (get a friend to help you)

1 Make sure the sea water has no seaweed, fish, or other things in it. To do this, place the filter in the filter holder and hold it over one of the glasses. Pour the water through the filter into that glass. (If you don't have sea water, stir a big spoonful of salt into a glass of tap water until all the grains disappear. You don't need to filter this water.)

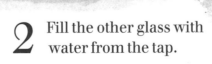

2 Fill the other glass with water from the tap.

3 Pour equal spoonfuls of powdered soap into both water glasses. Stir each glass until you can't see the soap powder any more.

4 Stick one straw into each glass. Give one glass to your friend.

5 Both of you blow through the straws at the same time, trying to make bubbles. Which glass has more bubbles? Which bubbles last longer?

water molecules

piece of soap molecule

pieces of salt molecules

WHAT HAPPENS?

You can blow a lot of bubbles with soap and fresh water, but salty water doesn't let you blow nearly as many. Fewer bubbles show that the water and soap molecules in that glass aren't sticking together well. In fact, it's even more difficult to make the soap powder completely disappear, or dissolve, in salt water.

When something dissolves in water, each of its molecules breaks into two pieces that stick to the water molecules. But there's a limit to the number of molecules that water can hold. There are already so many salt molecules dissolved in sea water that when you add soap powder, the soap molecules can't dissolve quickly. Pieces of salt molecules are already sitting in spots where the soap would go. Like kids who come late to the cafeteria, the soap molecules can't find a seat. And if the soap can't grab on to water, it can't mix with it to make bubbles or make you clean.

UNDER CONTROL

If this is an experiment about salt water, why do you test a glass of fresh water? Because you have to know how regular water reacts to soap before you can tell if salt water reacts differently. The part of an experiment that shows you how things regularly behave is called the control.

Here's why a control is important. Imagine that you're testing a new cat shampoo on 40 cats. If you shampoo all 40 cats and discover that 30 have silky coats, you'd probably decide the shampoo is great! To really test it, however, you should shampoo 20 cats, leave 20 alone, and then compare the two groups. If 15 of the shampooed cats have silky fur, but so do 15 of the unwashed cats, you've discovered that the shampoo makes no difference.

Why we wash

Do you wash your hands before eating?
Here's a way to see why getting dirt off your hands
with soap and water is a healthy idea.

You'll need:

- 15 mL (1 tablespoon) instant mashed potatoes
- 2 mL (1/2 teaspoon) sugar
- 5 mL (1 teaspoon) water
- a spoon
- a big jar lid you don't need any more
- a nail file
- dirty hands
- a small plastic bag

1 Mix the instant potatoes, sugar, and water in the jar lid. Stir them up into mush. Spread the mush smoothly over the bottom of the lid.

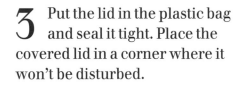

2 Scrape out your dirty fingernails with a nail file, letting whatever comes out fall on top of the mush.

3 Put the lid in the plastic bag and seal it tight. Place the covered lid in a corner where it won't be disturbed.

4 Look at the lid every day for a week. What do you see?

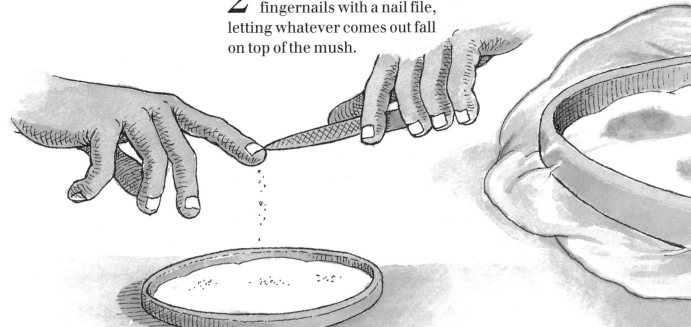

WHAT HAPPENS?

After five or six days, do you see dark fuzzy patches on the potato mush? Disgusting! Where did that fuzz come from? Actually, it grew from the stuff that was under your fingernails. Ick! Even more disgusting!

The dirt on your hands is full of tiny mold spores, "seeds" that grow into mold. (Mold spores are everywhere — even in the air around you.) Each of those spores can make more mold if it lands on the right food. Mold especially likes instant mashed potatoes, sugar, and water because they contain the basic foods that molds need to grow. Eventually there's enough mold on the mush that you can't help seeing it. Throw the bag out without opening it, unless you want even more mold in your house!

There are other tiny creatures on your dirty hands, too — bacteria, protozoa, and viruses. You can't see them, even when there are a lot, because they're even smaller than mold. But some of them can make you sick if they get inside your body. That's why it's smart to wash your hands before putting them on your food or near your mouth.

Does mold grow as quickly on mush mixed with "antibacterial soap"?

Does what's under your toenails grow molds different from what's under your fingernails?

STAY HEALTHY

Since your neighbor wasn't in school today, you've gone to his house to deliver an important homework assignment. You find him in bed, looking sicker than a gross joke. He warns you that he's "catching" — and you know he doesn't mean fly balls. So what can you do to keep from getting the same disease?

The best way to keep viruses and other germs out of your system is to wash your hands. Flu viruses are transmitted in droplets from someone's mouth or runny nose. If you have one of these viruses on your hand and then touch your nose, mouth, or even eye, the virus can get inside your body. Then you might catch the flu, too. Soap removes germs as well as dirt, so wash your hands! Some doctors estimate that they wash their hands twice for every patient they see: once before and once after.

9

Break the tension

If you think soap cleans your skin,
you should see it clean the "skin" off a bowl of milk!

You'll need:

- a scrap of tissue
- a deep dish or pie tin of milk
- a needle
- a clean spoon
- cinnamon
- blue food coloring
- a bar of soap

1 Drop the tissue onto the surface of the milk. Carefully lay the needle flat on the tissue.

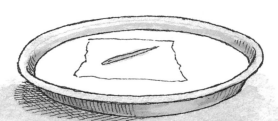

2 Use the spoon to gently push the edges of the tissue to the bottom of the bowl. The tissue sinks, but the needle stays on the surface!

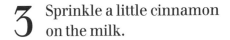

3 Sprinkle a little cinnamon on the milk.

4 Drip one drop of food coloring on the milk. Don't stir the liquids together.

5 Dip one corner of the soap into the middle of the bowl. What happens to the cinnamon, the food coloring, and the needle?

10

WHAT HAPPENS?

The needle looks as if it's floating on the milk, the cinnamon just sits there, and the food coloring sticks together in a blue drop — until you add soap. Suddenly the cinnamon zips away from the soap! The blue dot spreads out in an arc! Soon the needle goes under!

How does soap make these things move? To understand, you have to know that the needle doesn't float in the milk in the same way that you float in a lake or pool. Instead, it rests on top of the milk in the same way that you lie on top of a water bed. Milk contains at least 87% water, and the top layer of its water molecules stick tight to each other, forming a stretchy skin for the needle to lie on. The scientific name for water molecules sticking to each other this way is surface tension.

Even a tiny touch of soap breaks the surface tension because the water molecules start to stick to the soap molecules instead of to each other. Watch the cinnamon and the food coloring to see the milk's skin pull away to the edges of the bowl. And without the skin to hold the needle, it can't stay up. Imagine lying on a water bed and having the rubber covering disappear — SPLOOSH!

How many needles can you float close to each other before they stretch the milk's skin too much to hold them?

What happens if you use cream, which has more fat, instead of milk? What about milk with no fat at all?

SKIMMERS

Here's another way to play with surface tension and soap. Fill a bowl with water, and stir in a squirt of dishwashing detergent (try not to make any bubbles). Scoop up a cup of the soapy water. Then rest a big spoon upside-down on the side of the bowl so the spoon's tip almost touches the water. Dribble water from the cup over the back of the spoon. Watch drops fall off the spoon and briefly skim across the water in the bowl.

These "skimmers" may look like bubbles, but bubbles are filled with air and these skimmers are made of soapy water all the way through. They disappear not by popping like bubbles, but by finally being absorbed into the water below. Soapy water makes skimmers because its surface tension is strong enough to form water into droplets, but not strong enough to immediately pull those droplets into the rest of the water.

Pioneer soap

If you lived on a farm in pioneer times, you might have helped your family make soap. Sound like fun? Actually, it was a hot, smelly, messy job.

Imagine that you're living on a farm far away from town 150 years ago. Your mother tells you it's time to make enough soap for the rest of the year. (Because soap-making was done near the house, it was usually managed by the mother of the family.) She has built a fire outside under her big iron pot. Your job is to bring the two important ingredients for soap.

The first ingredient is meat fat. Your family has saved leftover fat for months, and since there are no refrigerators to store it in, it's now very smelly. Hold your nose as you lug out the tub full of fat. Your mother dumps it into her pot. Heating and stirring the fat for several hours will turn it into a smooth, thick liquid.

The second ingredient of soap is lye, which is a chemical found in the ashes of oak, maple, or birch wood. For months your family has swept the ashes from your fireplace into a barrel with a spigot at the

12

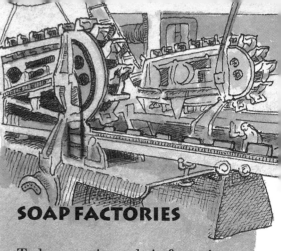

bottom. You poured water over the ashes, letting them soak until enough lye was dissolved into the water to float an egg. You open the spigot and fill a big bucket with lye water. Then you carry the bucket out to the yard. Be careful! Lye water stings if it touches your skin.

You and your mother take turns stirring the pot until the fat is finally smooth. Then your mother pours the top layer of fat into a wooden tub and adds lye water. The lye makes the mixture heat up as your mother stirs it. Gradually the molecules of lye and fat combine to form new soap molecules. After a few more hours of stirring, the soap becomes smooth and thick.

You and your mother tilt the tub and pour the warm soap into wooden boxes. It takes a whole day to cool and harden. At last your family has slabs of homemade soap for laundry, dish washing, and baths. After all that work, doesn't a nice long bath sound good? All you have to do is haul the water from the stream!

SOAP FACTORIES

Today soap is made in factories, but the science of soap-making is the same as it's always been. Fat is mixed with lye, causing a chemical reaction that produces warm soap. The warm soap is poured into molds to cool. The big difference is that factories make a lot more soap than one family could. The ingredients are mixed in huge vats with large mechanical stirrers, and the soap is poured into hundreds of molds at one time.

Soap factories also add extra ingredients to some types of soap. For instance, deodorant soaps are made to stop body odor. (That aroma is really caused by bacteria that live on everybody's skin.) Deodorant soaps work in two ways. Some, called antibacterial soaps, kill the bacteria that create body odor. Others contain perfumes that are supposed to smell clean and fresh. But what do you think "fresh" smells like?

13

Detergent: synthetic soap

You can use soap or detergent in most experiments — but are they the same?

You'll need:

- 2 identical jars with tight lids
- water
- a refrigerator
- 15 mL (1 tablespoon) laundry detergent powder
- 15 mL (1 tablespoon) soap powder
- a friend

1 Fill both jars with water, and leave them in the refrigerator for an hour.

3 Pick up one jar, and have your friend take the other. Shake them up. Try to make the powders disappear as they dissolve into the water.

4 Every couple of minutes, stop shaking and compare. In which jar can you see more grains of powder? Which has more bubbles?

2 Take the jars out of the fridge. Put the detergent in one jar and the soap in the other. Screw the lids on tight.

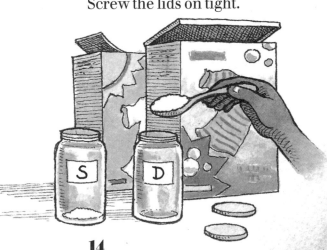

WHAT HAPPENS?

The detergent disappears quickly in cold water, and it makes lots of bubbles. Soap doesn't dissolve as easily in cold water, so it can't make many bubbles or clean clothes in cold water. (Lack of bubbles is a sign that soap doesn't work well in salty water, either — see page 6.)

Detergents are chemicals that stick to water and oil molecules in the same way soap does. But detergents can work better than soap: they grab oil very tightly and dissolve even in cold water. (Laundry powder also contains chemicals called builders that help detergents dissolve even more easily.) Because detergents are more powerful than soap, we now use detergents instead of soap to clean our clothes, dishes, and floors. Some soap companies even add detergents to bars of "soap."

How quickly do detergent and soap dissolve in warm water?
How does liquid soap compare to liquid laundry detergent?

STRONGER THAN SOAP

The first detergent was invented in Germany during World War I when there was a meat shortage. That meant there was also a shortage of fat for making soap. People needed a cleaner that could be made without fat but wiped off dirt just like soap. (The word detergent comes from a Latin word meaning "wipe off.") Since then scientists have invented hundreds of different detergent formulas.

Some detergents are so strong that they can't be used on people or clothing, only on metal machinery. Even dishwashing detergent can irritate people's skin. If your family washes dishes by hand, you might know about "dishpan hands": dry, painful patches of skin on hard-working hands. Dishpan hands are a sign to switch to milder soap, or to use rubber gloves.

Dishpan hands are also a good excuse to stop washing the dishes!

Floating soap

What's inside some bars of soap that makes them float? Here's one way to cook up the answer.

You'll need:

- a big bar of floating soap
- a plate that's safe to use in a microwave oven
- a microwave oven and permission from an adult to use it

1 Place the soap on the plate and center it in the microwave oven.

2 Set the microwave oven to heat on High for 45 seconds.

:45

:40

3 Watch the bar of soap through the door window. How do you see the soap reacting to the microwaves?

16

WHAT HAPPENS?

After about half a minute, the soap bar starts to swell. Its outside cracks! Soap oozes onto the plate like white slime! After the microwave stops, use a hot pad to remove the plate from the oven. Hold your hand over the soap to feel how warm it is. What's going on inside that bar?

Remember — a hot plate looks the same as a cold plate. But it feels a lot more painful!

The microwave oven produces microwave energy, which heats water molecules inside the bar. The hot water molecules then heat the tiny bubbles of air that are in the soap. As the water and air warm up, they expand, bursting open the outside of the soap and pushing lather out onto the plate. Those same tiny bubbles of air also make floating soap light enough to float, while a solid soap bar of the same size is heavier and sinks.

What can you do with your exploded soap bar? After it cools, scrape off the dried soap slime. It's too crumbly to wash with, but you can mix it with water to make bubble bath, or add it to other soap scraps to make your own bar of soap (see pages 18–19). The rest of the soap bar works perfectly well in the bathtub, even if it has as many holes as a Swiss cheese.

With your parent's permission, microwave a bar of soap that doesn't float. How does it react?

With your parent's permission, microwave a bar of moisturizing soap, which contains extra oils. Is it slimier than floating soap?

LUCKY LONG LUNCH

Sometimes inventors just get lucky — that's how Harley Procter and James Gamble discovered their floating soap, Ivory. It was ordinary soap until one day in 1878 when workers at the Procter & Gamble factory in Cincinnati accidentally left on the soap-stirring machine when they went to lunch. When they returned to the factory, they found that extra air had been mixed into the soap. They poured it into molds anyway, thinking nobody would notice.

Soon customers were asking for more floating soap. Procter & Gamble realized that their workers had invented a new product. They started to advertise Ivory as the soap that floats.

What's white and floats and cleans things?

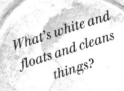

A polar bear with a vacuum cleaner!

Make a bar of soap

This recipe lets you mold your own soap bars — and recycle soap scraps left over from baths, science experiments, and soap carving.

You'll need:

- small plastic food container
- petroleum jelly
- 250 mL (1 cup) or more of soap scraps
- a cheese grater
- a measuring cup
- a steel or glass double boiler, with water in the lower pot
- water
- a plastic stirring spoon
- a stove, and permission from an adult to use it
- an adult to watch and help
- a chopstick

1 Coat the inside of the container with petroleum jelly. This will be the mold for your soap bar. Set it aside.

2 Grate the soap until the biggest scraps are about the size of a fingernail. Measure the soap scraps in a measuring cup, then put the soap into the top of the double boiler.

3 Add about half as much water as you have soap. For instance, if you have 500 mL (2 cups) of soap, add 250 mL (1 cup) of water.

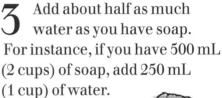

4 Turn the stove on to Medium heat. Stir the soap and water until the slivers melt into a smooth liquid.

5 As the water boils away, the liquid will thicken. Keep stirring every so often. It can take 30 minutes or more to melt the soap. Invite the adult to take turns.

6 When the melted soap has become as thick as honey, stand the chopstick up in the pot. If the soap holds the stick up, it's ready to pour.

18

7 Carefully pour your warm soap into the mold you prepared. Put it aside to harden.

8 The next day, turn the mold over and slip out your own bar of soap.

WHAT HAPPENS?

The soap slivers turn to soapy liquid in two ways. First, the water dissolves some of the soap. Next, the stove heats the water inside the double boiler, and that heat is carried into the soap-and-water mixture, melting the rest of the soap. The smaller you make your soap slivers, the quicker the water and heat turn them into liquid.

As you continue heating the soap, the water in the mixture turns from liquid to gas, which rises out of the pot. The melted soap becomes thicker and thicker. Eventually it's so thick that it will turn solid as it cools. That's when you pour the soap into your mold.

Make molds out of a glass, an old teacup, or a cardboard milk carton — whatever you like. Just be sure to choose either a container with a top wide enough for you to slip the soap out after it hardens or one that you can rip away when the soap's hard.

If you stir your melted soap a lot, does it make floating soap?

SOAP CARVING

Carving sculptures out of soap bars is a fun, artistic hobby. Harry Pierpont and Charley Makley had plenty of time to learn soap carving in 1934 — they were in prison for robbing banks. One day they showed the guards a gun and ordered them to open the cell door. The pair made it to the prison courtyard before being shot by other guards. Then the prison warden discovered the "gun" wasn't real: Pierpont and Makley had carved it out of soap and painted it black with shoe polish!

Why did the bank robbers bring along a bar of soap?

They wanted to make a clean getaway!

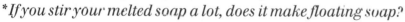

19

SOAP BUBBLE SCIENCE

You've played with soap bubbles many times before: blowing bubbles in the backyard, working up a froth in the sink, building mounds of suds in the tub. Did you know that playing with bubbles can be scientific?

The science of soap bubbles examines how bubbles are made and how they behave. A soap bubble is just a very thin layer of soapy water (called a soap film) wrapped around air. Bubbles are such fun to watch because soap films are so light and bouncy. But watch closely, because they can disappear in an instant!

To try out some bubbly science, all you need is soapy water, a frame to hold soap films, and air. This section tells you how to make the best soapy water, the biggest frames, and the most beautiful bubbles.

20

The best Bubble Brew

How do you make soap films? Soap in water is the solution — plus an extra ingredient to give your soap films extra staying power.

You'll need:

- 1 L (1 quart) water
- a clean 2 L (2 quart) bottle with a tight lid
- 100 mL (4 ounces) liquid dishwashing soap or detergent
- a measuring cup
- 45 mL (3 tablespoons) glycerine (in a drugstore's skin-care section — also called glycerol)
- a bowl

1 Put the water into the bottle.

2 Add the dishwashing soap. The easy way to get all the soap out of the measuring cup is to pour water from the bottle into the cup, swish the liquids together, and pour them both into the bottle.

3 Add the glycerine.

4 Cap the bottle tightly and shake the liquids up together. How does this mixture react to being shaken with air inside the bottle?

WHAT HAPPENS?

Already you can see bubbles form in this Bubble Brew — which shows soap films form easily from this liquid. Let the bubbles settle down. Soap films form best when the Bubble Brew is completely smooth, which means no froth or bubbles on its surface.

Dishwashing soap and water alone make fine suds, but soap films made from this Bubble Brew last longer because it also contains glycerine. The glycerine holds the water molecules in the soap film, which stops the film from drying out quickly and popping.

This recipe makes enough solution for three or four good bubbling sessions with the bubble blowers and other experiments in this book. You can always make more by mixing dishwashing soap, ten times as much water as soap, and a little glycerine.

Make the recipe above without the glycerine, and dip the Hoop (see page 32) in it. Do soap films made from this solution last as long as soap films made from Bubble Brew?

21

Bubble blower #1: the Loop

This basic bubble blower is simple to make — in fact, it's so simple you can buy it in stores.

You'll need:

- thin, flexible wire
- a bowl of Bubble Brew (see page 21)
- duct tape

1 Bend the wire as shown. Close the loop firmly.

2 Dip the Loop into your Bubble Brew, and blow gently on the film in the hole. Or, to save your breath, wave the Loop in the air.

3 If you have a hard time gripping your Loop after it gets soapy, make a thicker handle by wrapping tape around the straight end of the wire.

WHAT HAPPENS?

All bubble blowers work by making a frame for a film of soap solution to stretch across. When you blow air against the soap film, it stretches out, wraps around your puff of air, and breaks away. There's your bubble! You can even make a Loop with your fingers. Dip your hand in Bubble Brew, make the OK sign, and pull your hand out. Blow gently through the circle made by your finger and thumb for a handmade bubble! *Bend your Loop into different shapes: triangles, crescents, scoops. What sort of bubbles do these shapes produce?*

Bubble blower #2: the Tube

You'll need:

- a plastic drinking straw
- scissors
- a bowl of Bubble Brew (see page 21)

This bubble blower is also simple to make, and it gives you more control over your bubbles.

1 Flatten one end of the straw, and cut up the middle 1 cm (½ inch).

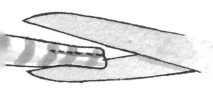

2 Flatten the two flaps, and cut up their middles the same distance.

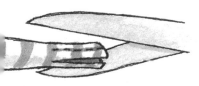

3 Bend all four flaps so they stick straight out.

4 Dip the end of the Tube with the flaps into your Bubble Brew.

5 Lift the Tube out of the Brew, and blow into the dry end. To detach the bubble, give the Tube a little twist.

WHAT HAPPENS?

The soap film on the Tube stretches across the narrow hole at its end. That film is much narrower than the film on a Loop, so how can it make an equally large bubble? Because the film in the Tube is thicker than the film on a Loop, and because there's extra Bubble Brew sticking to the flaps of the Tube. A long Tube channels the air from your mouth and makes it flow smoothly, which means the bubbles can grow larger.

Bubble blowout

You may have blown a lot of bubbles, but have you ever seen a bubble blow back?

You'll need:

- a funnel
- a bowl of Bubble Brew (see page 21)
- a lit birthday candle

1 Wet the whole funnel with Bubble Brew, but wipe off the narrow end.

2 Dip the wide end of the funnel into the bowl. Let extra Bubble Brew drip off for a couple of seconds.

3 Blow steadily through the narrow end of the funnel until you make a bubble about twice as wide as the funnel's wide end.

4 Take the funnel out of your mouth, and point its narrow end at the candle.

WHAT HAPPENS?

As the bubble shrinks, it pushes air out through the small hole in the funnel. The candle goes out because that moving air carries away the fuel it needs to burn: the cloud of melted wax around its wick. The bubble blows out the candle in the same way you could blow it out with your breath. In fact, the air coming from the bubble is your breath.

Why does the bubble shrink? Soap films always try to shrink because the water molecules inside them are being tugged closer to each other by surface tension. When the sides of a soap film are held by a frame, the film fits into that frame in whatever shape has the smallest surface. A film held by a funnel, for instance, shrinks until it reaches the narrow end, where it has the smallest space to stretch across.

Try this experiment with different sizes of funnels. Can a small bubble blow out the candle? What about a bigger bubble?

Is a bubble on a funnel strong enough to spin a pinwheel?

FIRE EXTINGUISHER

A soap bubble can put out a small flame, but bubbles of another kind can put out huge fires. When fire fighters are faced with a jet-fuel fire, they spray a foamy fire-extinguishing chemical that looks like soap lather onto the flames. The foam works in two ways: it smothers the flames so they can't get enough oxygen to spread, and it cools whatever's burning. If fire fighters threw water on these fires, the jet fuel would float on top of the water (just like cooking oil) and the flames would spread.

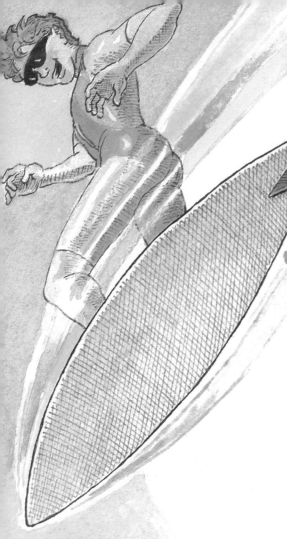

Inside a bubble

Hey, dudes! You want to catch a wave on a soap film? Coasting down the inside of a bubble will just blow you away, so grab your boards.

Whoa! Before you touch the soap film, you have to totally cover yourself in soapy water. I hope you brought a wet suit.

Here we are on the outside of the bubble — or, as we soap-surfers call it, the outside. We're hanging ten over a thin layer of dishwashing soap dissolved in water. This particular film is about a millionth of a meter thick; that's more than 50 times thinner than one page in this book. Radical, huh? Now, while I've been talking, we've started sinking into the soap film. It's a good idea to hold your breath, dude.

You need soapy water to make a bubble because pure water has too much surface tension to stretch out into a film. The soap helps it chill out and lose some of that tension. Once you've got a stretchy soap film, you can blow on it. The film curls around that puff of air to make a most excellent bubble.

This bubble was blown by someone doing soap science. (Someone who just had peanut butter for lunch — gnarly!) The air in here feels a little stuffy, doesn't it? The air inside a bubble has to be a little thicker than the air outside because it's under extra pressure from the soap film trying to shrink.

Cowabunga! We can shoot an awesome curl here inside the bubble because its walls are perfectly spherical — or, as we hot doggers like to say, round. They're round because soap film always

pulls itself into the shape with the smallest surface area. Surface area is the amount of outside — get it? And the shape with the smallest surface area happens to be a sphere. Like, the same volume of air inside this bubble could be shaped like a box or a tube or a football, but to make all those shapes, the soap film would have to stretch more.

Bummer! The waves are dying down. The soapy water is flowing to the bottom, so the top of the bubble is thinning out. It won't last long in this hot sun. After a while the top's going to get so thin that the soap film won't stretch any more, and then you know what'll happen?

BOOOOM! Wipe-out, dude!

Bubbles burst at speeds over 50 km/h (30 miles per hour)! Very thin soap films peel back more than twice as fast as that!

27

Framed in 3-D

How do soap films behave when they stretch across three-dimensional frames? Make two frames, and you'll know all the angles!

You'll need:

- a plastic drinking straw
- 2 strings, each 20 cm (8 inches) long
- a clean cardboard milk carton
- scissors
- a big bowl of Bubble Brew (see page 21)

1 Form the straw into a triangle by bending it in three places and sticking one end into the other.

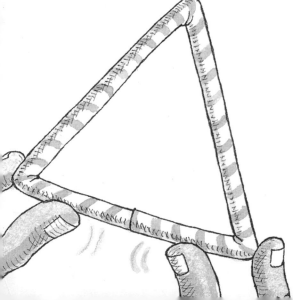

2 Tie each end of one string to a different corner of the triangle.

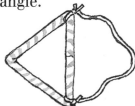

3 Tie one end of the other string to the third corner. At that string's halfway point, knot it around the middle of the loop formed by the other string.

4 Lift the end of the loose string. Do you see how the strings create three more triangles with the straw below? This shape made from four triangular sides is called a tetrahedron. Put it aside.

5 Cut the top and bottom off the milk carton so you have a shape that's as tall as it is wide.

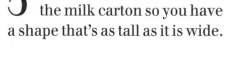

6 Cut out the middle of each side of the carton, leaving 1 cm (1/2 inch) of cardboard at the top, bottom, and right side of each panel. This frame is a cube.

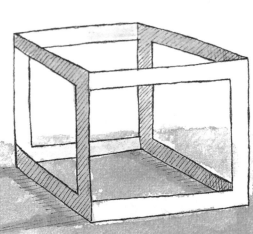

28

7 Guess how soap films will stretch across the cube and the tetrahedron. Then dip each frame in the Bubble Brew. What do the soap films look like?

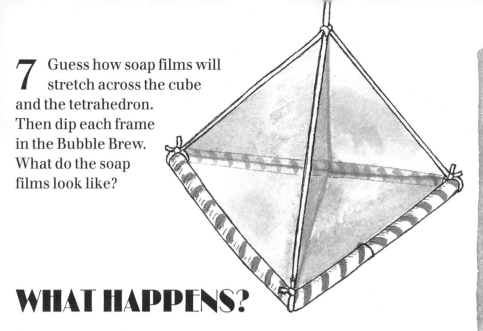

WHAT HAPPENS?

Instead of covering the outside of your frames like windows, the soap films reach into the middle of the frames and meet there. That's because soap films always pull themselves into the shape with the smallest possible surface area.

On the tetrahedron, soap films make six small triangles within the frame instead of four big triangles on its outside. The six surfaces inside the tetrahedron add up to only about 60% of the frame's outside area.

On the cube, soap films make triangles, trapezoids (four-sided tapering shapes), and a rectangle in the middle. Does the rectangle always face the same side of the frame?

Pop a triangle inside the tetrahedron. How does the soap film react?
Pop one of the surfaces inside the cube. How does the soap film react? Does it react the same way if you pop a surface with a different shape?

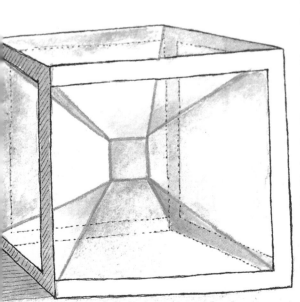

Wet a Tube (see page 23) with

BUBBLE CUBE

When three soap films meet at a corner, they always form angles of 120°. When four films meet, the angles between them always measure 109°28'. Those two angles stretch soap films the least. But you can still blow a bubble shaped like a cube, whose angles are all 90°. Dip your cube frame into Bubble Brew so it's full of soap films. Wet a Tube (see page 23) with soap solution, touch the end to the rectangle at the center of the frame, and blow gently through the tube. The bubble that grows out of that rectangle is shoved into a cube shape by the soap films all around it.

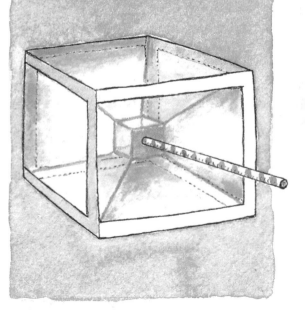

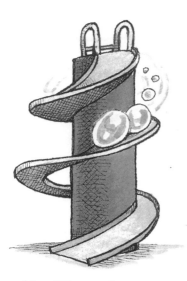

Spiral soap film

Have you ever slid down a slide that twists around and around like a corkscrew? Here's a different kind of twisting slide — one that's made out of soap film.

You'll need:

- a flat plastic lid from a food container
- scissors
- a stapler
- 20 cm (8 inches) of string
- a saucer of Bubble Brew (see page 21)

1 Cut a spiral in the plastic lid, cutting about 1 cm (½ inch) closer to the middle each time around, as shown.

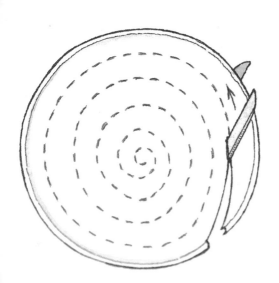

2 Fold the outside tail of the spiral at a 90° angle so that the end lies under the center of the spiral. Staple the fold so it stays in place.

3 Staple one end of the string to the tail at the bottom of the spiral. Staple the center of the spiral to the string, leaving about 5 cm (2 inches) of string hanging loose.

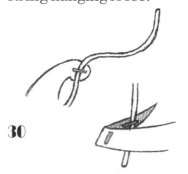

4 Pick up the spiral using that loose string as a handle. Dip the whole thing into Bubble Brew, and pull it out gently. What shape does the soap film take?

WHAT HAPPENS?

The soap solution stretches between the plastic twirl and the string to create a screw shape. How many soap films do you see? Only one. Take a close look — the screw is made of one continuous film that twists back over itself.

This soapy slide won't last forever, unfortunately. Gravity pulls the soap solution towards the bottom, and the film dries out. If you look closely, you can see swirls appear as the liquid flows down the screw. Soon the top of the film is too thin to stay stretched, and — poof — the soapy slide is gone!

Build more spirals with longer strings and wider lids. How far can a soap spiral stretch?

Dribble a drop of Bubble Brew onto the top layer of the soap screw. How many layers does the drop pass through?

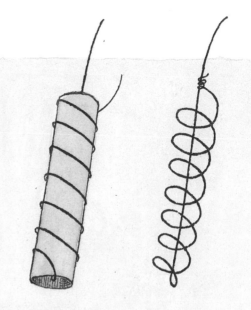

WIRE SPIRAL

Here's another frame for soap spirals. Take 1.5 m (5 feet) of thin, flexible wire. Poke it up the inside of a cardboard tube 30 cm (1 foot) long until about 10 cm (4 inches) of wire sticks out the top. Twist the longer end of the wire around the tube, leaving about 3 cm (1 inch) between each twist. When you get to the top of the tube, slip the wire off the tube. Wrap the ends of the wire together, and dip your frame in Bubble Brew.

You can also build a cube and a tetrahedron from wire. Or twist out other shapes: prisms, pyramids, whatchamacallits! How does soap film behave when you dip them?

31

How many frothy bubbles can you pop with a dry finger?

Only one — then your finger's not dry any more!

Bubble blower #3: the Hoop

The Hoop produces bubbles so BIG that you should use it only outside — or mop the floor afterwards.

You'll need:

- a wire clothes hanger
- 1 m (3 feet) of yarn
- duct tape
- a dishpan of Bubble Brew (see page 21)

1 Bend the loop of the clothes hanger into a circle. It should be about 20 cm (8 inches) across.

2 The hanger's hook is your handle. Wrap it in tape so there are no ends to poke you.

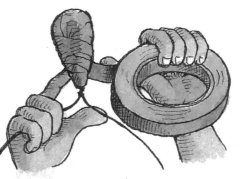

3 Wrap yarn around the outside of the Hoop so that it looks like the stripe on a candy cane. Don't miss the area where the handle joins the loop.

4 Dip the Hoop in Bubble Brew and pull it out. Is there a film across the hole? If the weather is bad for bubbles (see the box), the frame may be too wide for a film to form. To make your Hoop smaller — about 15 cm (6 inches) across — use pliers to twist another loop near the handle.

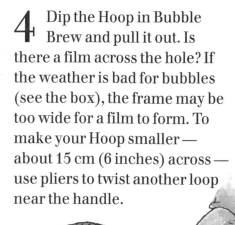

5 When there's a soap film across the Hoop, let extra solution drip off for five seconds. Then wave the Hoop in slow loops through the air. How big are these bubbles?

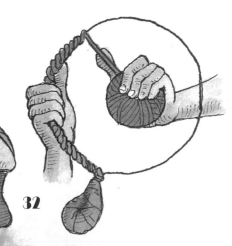

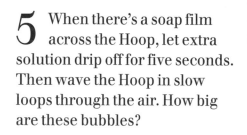

32

WHAT HAPPENS?

The Hoop works just like a giant Loop (see page 22): because it holds a large soap film, it makes b-i-i-i-g bubbles. The yarn helps by absorbing extra soap solution and gripping the film, so you have more to build bubbles with. Unfortunately, the yarn also produces froth — lots of little bubbles that break up large soap films. If you see froth in the dishpan, clear it away with a dry finger.

The Hoop is great for making one bubble around another. Start by blowing a small bubble with your Loop. Then try to "catch" that bubble in the air with the Hoop, as if you're netting a butterfly or playing tennis very slowly. If you've got good aim, you can catch the small bubble inside a big one. Twist your wrist to send them floating off together!

How well does a Hoop work without the yarn?
Can you make bubbles by blowing through the soap film on a Hoop?

(see page 22)

GOOD BUBBLE WEATHER

"This is meteorologist Bubbles Bellaire with today's weather report. It's a beautiful morning for making bubbles! There's no chance of rain, but the hygrometer measures our humidity at a near-perfect 90%! That means there's enough water in the air for bubbles not to dry out quickly. Of course, people will feel pretty sticky.

"This afternoon the wind will pick up a little. Strong gusts mean you'll have trouble making a big bubble that stays together. Make medium-sized bubbles, or chase small bubbles through the air.

"Just a reminder: Don't make bubbles over a nice green lawn. Too much soap can harm the grass. The best surface for big bubble-making is an empty parking lot."

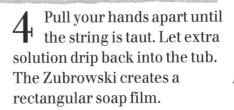
What's the largest Zubrowski you can make?

Bubble blower #4: the Zubrowski

This flexible bubble frame was invented by Bernie Zubrowski at The Children's Museum in Boston.

You'll need:

- 2 plastic drinking straws
- 1 m (3 feet) of string
- a dishpan of Bubble Brew (see page 21)

1 Drop the string down through each straw. (If you have trouble, use a knitting needle or get an adult to help poke the string through.)

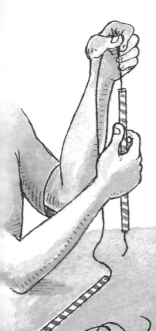

2 Tie the ends of the string together. Pull the knot into one of the straws so the frame has four smooth sides: two string, two plastic.

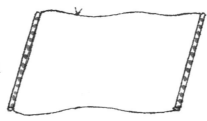

3 Hold the Zubrowski by the straws. Dip it into the Bubble Brew, pulling it out with the straws about a hand's width apart.

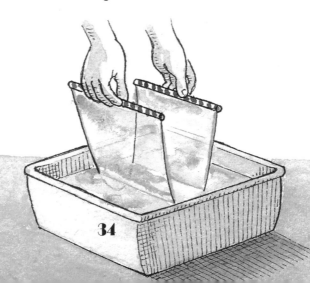

4 Pull your hands apart until the string is taut. Let extra solution drip back into the tub. The Zubrowski creates a rectangular soap film.

5 You can make bubbles with the Zubrowski in three ways: For small bubbles, hold the Zubrowski by opposite sides of the loop of string and stretch it out. You should have a very long, narrow film of soap. Blow gently while moving the film from right to left in front of your mouth. With practice you can blow a stream of small bubbles.

For medium-sized bubbles, blow into the middle of the soap film. Usually a bubble forms and drifts away, and the film closes up again. Try to make a medium-sized bubble and catch it in a BIG bubble.

For BIG bubbles, hold your arms apart and swing the Zubrowski through the air. A bubble starts to form out the back end of the Zubrowski. To set it free, bring your hands together. In the right conditions, these bubbles can grow as large as beachballs.

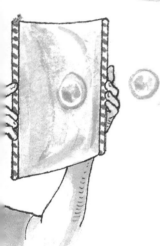

WHAT HAPPENS?

You have to tug to keep the straws of the Zubrowski apart! The soap film is always pulling on the strings, trying to shrink as small as it can. Let the Zubrowski hang by one straw and you'll see the soap film's pull. That hourglass shape is created by the balance of two forces: the shrinking soap film, and gravity pulling on the bottom straw. Pull down on that straw, let go, then watch the soap film lift the straw.
What happens if you twist the bottom straw around so far that the strings touch in the middle?

BOUNCE A BUBBLE

Soap film is so elastic that you can use it to build a trampoline for bubbles. Dip your Zubrowski in Bubble Brew and hold it out like a tray. Then have a friend blow a bubble from the Loop (see page 22) into the air in front of you. Move the Zubrowski under the bubble, and gently move your arms upward. It may take some trying, but you can get the bubble to bounce on the sheet. How many times can you bounce a bubble? What happens if the bubble lands on one of the strings?

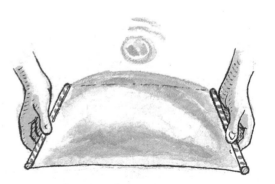

35

BUBBLE MYSTERIES

"Bubble mysteries?" you say to yourself. "What's so mysterious about bubbles? I blow bubbles out of soap film. Some are big, some are small, and they all float around and pop. I don't see anything mysterious about that!"

Bubble mysteries aren't what you see — they're what you can't see. The experiments in this section deal with stray bits of electricity, invisible gases, heat waves, and more. These scientific marvels are around us all the time, but we can rarely see them. Adding bubbles makes all these mysteries visible.

The bubble electric

**Have bubbles always attracted you?
Here's how to attract a bubble.**

You'll need:

- a heavy book
- a table
- a cup of Bubble Brew (see page 21)
- the Loop (see page 22)
- an inflated balloon
- a head of clean, dry hair
  (preferably your own)

1 Lay the book flat on the edge of a table.

2 Blow a bubble and catch it on the Loop. Stick the Loop handle under the book so the bubble hangs off the table.

3 Rub the balloon on your hair. If the air is dry, you might even hear it crackle.

4 Hold the balloon near the bubble. Try moving the balloon closer and farther away.

WHAT HAPPENS?

Move the balloon to the right, and the bubble reaches right. Move the balloon left, and the bubble reaches left — maybe even so far that it pops. What attracts the bubble? A little something from your hair.

Your hair is made of molecules, and molecules are made of atoms. And atoms are made of particles called protons and electrons. Protons have a positive electrical charge (+) and stick together in the middle of atoms. Electrons have a negative charge (-) and flit around the outside of atoms, so they come loose easily. Rubbing a balloon on your hair knocks electrons off your head onto the balloon.

How do extra electrons on the balloon make the bubble stretch? Electrons push away other electrons, but they attract protons. When the electron-carrying balloon gets near the bubble, its electrical charge pushes loose electrons in the soap film to the far side of the bubble. That leaves the bubble's near side with extra protons, which are attracted to the electrons on the balloon. That attraction is strong enough to make a light, flexible bubble move.

Grow a bubble

Have you been blowing so many bubbles that you're out of breath? Now you can watch a bubble blow up by itself.

You'll need:

- 15 mL (1 tablespoon) baking soda
- a spoon
- a glass jar with a wide mouth
- 125 mL (1/2 cup) vinegar
- the Loop (see page 22)
- a cup of Bubble Brew (see page 21)

1 Spoon the baking soda onto the bottom of the jar.

2 Slowly pour the vinegar onto the baking soda. The mixture starts to fizz and bubble.

3 While you wait for the fizzing to calm down, use the Loop and Bubble Brew to blow some bubbles, and catch one on the Loop.

4 Carefully lower the bubble into the jar and hold it there for a minute. How does the bubble change?

WHAT HAPPENS?

After you lower your soap bubble into the jar, it starts to grow — so big that it could even pop! What's in the jar that can blow up a bubble, and how can it get inside the bubble without opening the soap film?

Mixing baking soda and vinegar produces a gas called carbon dioxide. Carbon dioxide dissolves in water, but the amount of gas that is dissolved always depends on the amount of gas in the air around the water. When there's more carbon dioxide in the air than in the water, the gas dissolves quickly. When there's less carbon dioxide in the air, those molecules leave the water and join the air again.

When you put your bubble into the jar, there's so much carbon dioxide around it that some of that gas dissolves into the water in the soap film. The dissolved molecules spread out within the film and some end up on the film's inner surface. Because there's less carbon dioxide in the air inside the bubble, those molecules leave the water and join the air there. More carbon dioxide inside the bubble makes it grow.

*What happens to a big bubble placed across the top of the jar?

*What happens to two joined bubbles planted in the jar?

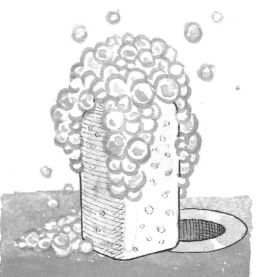

BUBBLE FOUNTAIN

Since carbon dioxide is invisible, you can't see it form — unless you've got a squirt of soap. Place a 1 L (1 quart) bottle in the sink — this can get messy! Swirl 500 mL (2 cups) of water and 15 mL (1 tablespoon) of baking soda together in the bottle. Stir a squirt of liquid soap into a measuring cup holding 125 mL (1/2 cup) of vinegar. Then pour the vinegar and soap into the bottle. Watch the bubbles cascade!

All those bubbles are filled with carbon dioxide created by the combination of the vinegar and the baking soda. They mix together fast and furiously. (Put your hand on the bottom of the bottle to feel the energy produced inside.) Carbon dioxide is created the same way inside your bubble-growing jar.

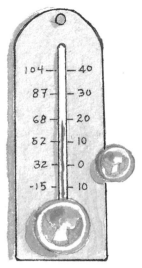

Bubble thermometer

Do you have a warm feeling about bubbles?
Or do they leave you cold? This bubble-in-a-bottle
can tell you whether it's feeling hot or cool.

You'll need:

- the Tube (see page 23)
- a cup of Bubble Brew
  (see page 21)
- a glass bottle with a narrow neck
- a bowl of hot water
- a bowl of cold water

1 Dip the Tube in the Bubble Brew, and then blow some bubbles inside the bottle. All you need is one soap film across the middle of the bottleneck.

2 Place the bottle in the bowl of hot water. Let it sit for a few seconds and watch how the film in the bottleneck reacts.

3 Take the bottle out of the hot water and put it in the cold. What does the soap film do now?

WHAT HAPPENS?

When air, or any other gas, gets warmer, it tries to fill more space. The warmer air inside the bottle expands and pushes the soap film up. When you move the bottle to the cold water, the air cools down. Cool gas needs less space than warm gas, so it pulls the soap film back down. Since the soap film in your bottle shows temperature changes, you've actually made a very simple thermometer. How many times can you send the film up and down before the bubble pops?

If you cap the bottle, expanding air at the bottom can't make room for itself by pushing air out the top. Does heat make the soap film rise in a capped bottle?

Blow a bubble that perches on the lip of the bottle. How does it react to heat and cold?

UNBELIEVABLE

"Let me show you the unbelievable psychic energy of my mind!" you tell your friend.

"*Your* mind?" your friend replies. "I don't believe it!"

"I'll give you a hands-on demonstration. Put your hands on the bottle the same way I do and watch that soap film in the neck."

Your friend grips the bottle. Then, right before your eyes, the film gradually starts to rise! Is it your psychic energy? No, it's your body energy! Depending on your blood circulation, the palms of your hands and your friend's hands are as warm as 22–32°C (70–90°F). Your heat warms the air inside the bottle, making it expand.

Test your lungs

Doctors call the amount of air that you can blow out in one breath your vital capacity. It's a good measure of how healthy your lungs are. You can check your vital capacity with a soap bubble.

You'll need:

- a flat tray of Bubble Brew (see page 21)
- a chair
- a Tube (see page 23) made with a wide straw
- a ruler

1 Fill the tray with Bubble Brew. Rub the edges with Bubble Brew, too, because dry areas will pop bubbles.

2 Sit in the chair. Take long breaths. Don't try to suck in all the air in the room! Take six long breaths and slowly fill your lungs with oxygen.

3 Put one end of the Tube into the Bubble Brew, and blow through the Tube softly and steadily. Raise the Tube to keep it on top of the bubble.

4 Stop blowing when you've exhaled all the air in your lungs — don't let your chest start to hurt! Close the bubble by quickly twisting the Tube and pulling it away.

5 Before the bubble pops, locate its edges on the tray. Measure the distance across the bubble.

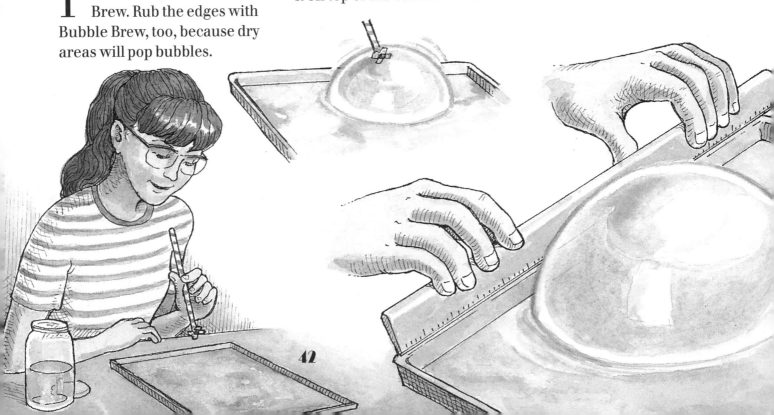

42

WHAT HAPPENS?

One thing that might happen during this experiment is that the bubble breaks. Big bubbles are fragile, so don't be discouraged. Take a few long, slow breaths to refill your lungs, and try again.

Measuring the distance across the bubble tells you approximately how much air came out of your lungs. That's because a bubble on a flat tray comes close to the shape called a hemisphere (half a sphere). It's close enough that you can use the mathematics of a hemisphere to find out the volume of air inside the bubble — and inside you.

First, divide the distance across the bubble in half. That number is called the hemisphere's radius.

Multiply the radius times the radius, times the radius again.

Multiply the number you have now times 2.

Divide the result by 3.

Finally, multiply what you have by pi (pronounced like "pie"), which is a number close to 3.14. It's often written as the Greek letter π. Pi is important for measuring round things, such as circles, bubbles, and pies.

SOAP FROM THE GARDEN

Several plants produce saponin, which Native people of North America once used as soap. Go on a field trip to find the most common kind of soap plant: a wildflower called soapwort. Look for patches of it in moist places between July and September. The flower is white or pinkish white, about 3 cm (1 inch) wide, with five folded-back petals. Squeeze the soapwort petals between your fingers and they'll ooze a liquid that feels slippery and creates a lather, like liquid soap. Then wash your hands with real soap because soapwort is slightly poisonous.

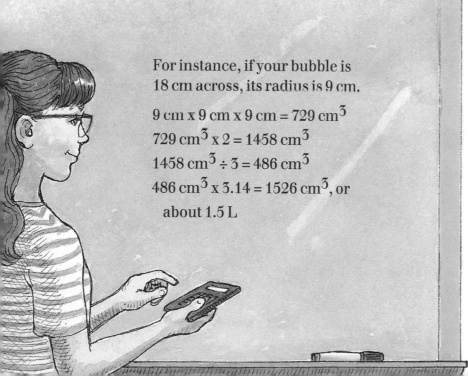

For instance, if your bubble is 18 cm across, its radius is 9 cm.

$9 \text{ cm} \times 9 \text{ cm} \times 9 \text{ cm} = 729 \text{ cm}^3$

$729 \text{ cm}^3 \times 2 = 1458 \text{ cm}^3$

$1458 \text{ cm}^3 \div 3 = 486 \text{ cm}^3$

$486 \text{ cm}^3 \times 3.14 = 1526 \text{ cm}^3$, or about 1.5 L

Bubbles and breezes

Air is moving all around you, even indoors, but you can't see it — unless you have help from a lot of little bubbles.

You'll need:

- a pan of water
- a stove and permission from an adult to use it
- the Tube (see page 23)
- a glass of Bubble Brew (see page 21)

1 Put the pan of water on the stove, and set it to boil.

2 Step away from the stove. Dip the Tube in the Bubble Brew, tap off the extra liquid, and practice blowing streams of small bubbles. These bubbles are your control (see page 7). Most of them probably head right for the floor.

3 Once the water has been boiling for a minute, blow more streams of small bubbles over the pan. Don't lean over the stove; just let your breath push the bubbles over the pan. Do any of these bubbles behave differently from your control? Do any of them rise?

4 Pick out one or two rising bubbles to watch. Can you track their paths through the air?

44

WHAT HAPPENS?

When the water hits 100°C (212°F) and boils, it sends up a current of steam (which is invisible), hot fog (which looks like silvery wisps), and warm air heated by the steam and fog. Many of the bubbles that fly into this current fall just like your control bubbles — they're already heavier than air, and fog drops add even more water to them. But a few are light enough that the current pushes them upward. These are the ones to watch to see the air currents in your kitchen.

Watch those light bubbles as they float through your kitchen, riding invisible air currents. You can see how the warm air over the pan rises. When the same air gets too far from the hot water, it cools, and the bubbles show you that cooler air falls. Sometimes a bubble moves up, down, or sideways even when it's not over the pan, showing how many little breezes are blowing through your house.

Do large bubbles float in the current? Try catching a large bubble on a Loop (see page 22) and, using tongs, holding it above the hot fog.

How do bubbles behave around your home's heater or air conditioner?

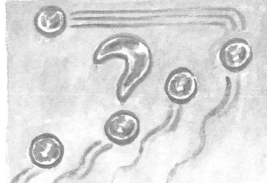

THE SAMPLE QUESTION

Do bubbles riding on air currents usually hit the ceiling? Find out by watching some. But be sure to watch a dozen bubbles or more — after all, the first couple might have caught a ride on an odd breeze. But if 12 out of 12 bubbles miss the ceiling, you can be very certain that the next one will miss. In science, the number of times you try one experiment is called a sample, and the bigger your sample, the more sure you can be of your answer.

After observing a dozen bubbles head upward, you can be mighty sure that these bubbles don't get closer to the ceiling than about 1 cm (½ inch). Just when it looks as if a bubble will land on the ceiling, it scoots to the side. That shows you there's an air cushion just below the ceiling, keeping the rising air current away.

Helium bubbles

Don't let your helium balloon float away! If it lands in the ocean, a whale might swallow it and get a whale of a stomachache. Instead, use the balloon to make helium bubbles.

You'll need:

- a glass that's less than 5 cm (2 inches) across at the top
- a cup of Bubble Brew (see page 21)
- a new helium balloon
- a clothespin or small clip
- a Tube about 10 cm (4 inches) long (see page 23)
- a rubber band

1 Fill the glass with Bubble Brew.

2 Twist the neck of the balloon above the knot, and pinch it closed with the clothespin. Seal the neck tightly to keep the helium inside.

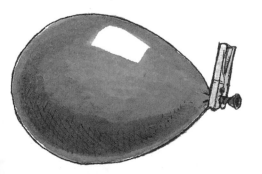

3 Carefully undo the balloon's knot. If helium starts to escape, twist the neck closed and seal it again with the clothespin.

4 Stick the Tube into the mouth of the balloon. Using the rubber band, tightly fasten the neck around the Tube. Remove the clothespin. By untwisting the neck of the balloon just a little, you can let out a slow, steady stream of helium.

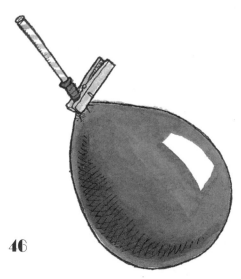

5 Stick the end of the Tube into the glass of Bubble Brew. Now you can make two kinds of helium bubbles:

For LARGE bubbles, leave the Tube in the Bubble Brew. Let out enough helium to make bubbles on the liquid's surface. Poke the Tube into the largest bubble and steadily let out more helium. When the bubble is as large as a plum, it breaks off and rises straight up.

For small helium bubbles, pull the Tube out of the Bubble Brew. Point the end of the Tube up in the air, then let out a little helium. A small bubble forms at the Tube's end. When it's the size of a big gumball, it breaks off and rises straight up.

WHAT HAPPENS?

Most bubbles float down and pop on the ground, but helium bubbles float up and pop on the ceiling. Even bubbles carried up by air currents don't get that far because the currents aren't strong enough (see pages 44–45). But helium gas is lighter than air, so helium always rises to the top of the air. Even with the weight of the soap film, these helium-filled bubbles are lighter than air.

This experiment works best with a new balloon. A rubber balloon that's more than a day old has already lost some of its helium. (Mylar balloons hold helium longer.) Use a narrow glass so the big bubble won't have much Bubble Brew to hang on to as it grows. Bubbles try to hold on to the liquid from which they grow, and the more of that surface they can touch, the tighter they stick.

Take the helium bubbles outside. How high can they rise?

SOAPBOX RACERS

Soap used to be shipped to stores in big wooden crates. Kids took those boxes, attached wheels from wagons or roller skates, and raced each other down hills. In 1933 a reporter in Dayton, Ohio, who was writing about the kids' cars named them soapbox racers and organized the All-American Soap Box Derby. The first winner was 16-year-old Randall Custer, and the runner-up was 11-year-old Alice Johnson. Then a car company started to sponsor the soapbox derby every year. Today there are regional meets all over North America, and the best racers compete for the annual championship in Akron, Ohio. The race cars aren't made of soapboxes any more; instead, kids build them from fiberglass and other space-age materials so they move through the air quickly.

Use the Loop and the Tube (see pages 22–23) to make a bunch of connected helium bubbles. Are they light enough to float?

47

Bubble colors

Sometimes soap films have no color, and sometimes they can show you every color of the rainbow.

You'll need:

- a flashlight
- a dark room
- the Hoop (see page 32)
- a bowl of Bubble Brew (see page 21)
- a glass bottle

1 Turn on the flashlight, go into the dark room, and set the flashlight down so it lights your work area.

2 Dip the Hoop in the Bubble Brew. Insert the handle of the Hoop into the bottle so the Hoop stands up and stays steady.

3 Shine your flashlight at the soap film and watch its reflection. Point the flashlight at the top of the film. What colors do you see?

4 Let the soap film drip for about half a minute. Shine the flashlight at the bottom. What colors do you see there? Can you see patterns?

48

5 Point the flashlight at the top of the soap film again. What does the reflected light look like now?

WHAT HAPPENS?

You shine white light on a clear soap film, and you see colors! Pink, green, yellow, blue, orange — where do they come from? Those colors come from the white light, which is a combination of all colors our eyes can see. Light travels in waves, and each color has a different shape of wave. When light waves of every color come out of the flashlight at once, the light looks white.

As you shine the flashlight at the soap film, most of the light passes through the Hoop and hits the wall behind. But about 4% of the light is reflected back by the front of the film, and 4% more is reflected by the back surface. Even a soap film has a front and back. So you have two sets of light waves bouncing back at you.

Sometimes a light wave from the front of the film interferes with the same color light wave from the back. For instance, the top of one pink light wave may line up with the bottom of another pink light wave. The two waves will cancel each other and you don't see *any* pink. Since the rest of the light bounces back, you see greenish white (every color but pink) instead.

Whether the two light waves cancel each other depends on their shape and how far apart the reflecting surfaces are. A soap film of a particular thickness always cancels one color of light. So by looking at the tints of the soap film, you can tell how thick it is.

Pink and green stripes mean the soap film is somewhere between 750- and 1500-billionths of a meter thick. Purple, blue, green, yellow, orange, and red swirls mean that the soap film is even thinner, somewhere between 200- and 750-billionths of a meter thick. The swirls appear because some parts of the film are trying to pull liquid from other parts. Finally, at the top of the soap film, you see only a dim, spotty reflection of the flashlight. That means the film is less than 30-billionths of a meter thick, and it's about to pop!

49

A LITTLE BIT OF SOAP

"Can someone hand me the soap? I'm in the shower and I can't find the — Hey! What's this little sliver in the soap dish?

"All right, who used almost the whole bar of soap and left nothing but this tiny sliver? This scrap of soap is small enough to go down the drain, even with the stopper in! A mouse couldn't even wash behind its ears with it! What good is a little bit of soap?"

The experiments in this section will show how powerful a little bit of soap can be. Just a scrap or a squirt can help you copy newspaper cartoons, change red juice to blue, or set up an electrical current.

Basic soap

Soap in your eyes — aaaaaagh!
Why do soapsuds hurt your eyes so much?

You'll need:

- 3 leaves of red cabbage
- a blender and permission from an adult to use it
- hot water
- a sieve
- a large glass or glass bowl
- powdered soap or detergent

1 Your first task is to turn the cabbage leaves into mush. Start by tearing the leaves into scraps about the size of a penny.

2 Put the scraps in the blender, and cover them with hot water. Cover the blender and purée the scraps.

3 Hold the sieve over the glass and pour the cabbage mush through it to fill the glass with the juice. Throw out the mush.

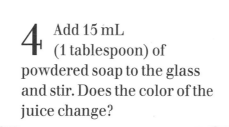

4 Add 15 mL (1 tablespoon) of powdered soap to the glass and stir. Does the color of the juice change?

WHAT HAPPENS?

Cabbage juice is an acid-base indicator. It turns more red when it's mixed with an acid, and more blue when mixed with a base. Since soap makes the cabbage juice turn bluish purple, soap must be a base.

Strong bases, like lye, can hurt your skin. Soap is only a mild base, but it still irritates the tender nerves around your eyes. That's why your eyes sting until you rinse the soap out. Soap in your mouth is no fun, either — all bases taste bitter.

51

Blue skies

Why is the sky blue? Why does the Sun look orange at sunset? And what on Earth do those questions have to do with soap?

You'll need:

- a stubby pencil
- a flashlight
- aluminum foil
- a piece of thick white paper
- a glass of water
- a dark room
- powdered soap
- a spoon

1 Hold the pencil against the flashlight lens so it sticks straight out, and wrap foil tightly around the pencil and the top of the flashlight.

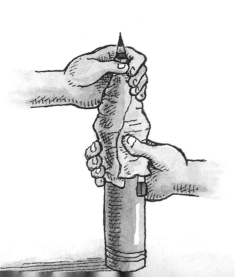

2 Remove the pencil, leaving a cone of foil wrapped around the flashlight and tapering to a hole no wider than 1 cm (1/2 inch).

3 Fold the paper in half so it can stand up as a screen behind and to the left side of the water glass. Turn off the lights in the room.

4 Shine the flashlight beam at the right side of the glass. What color do you see in the glass? What color do you see on the paper to the left?

5 Stir a heaping spoonful of powdered soap into the glass so the water turns cloudy.

6 Shine the flashlight beam at the right side of the glass again. Do you see a streak of light in the glass? What color is it? What color do you see on the paper to the left?

WHAT HAPPENS?

Light shining through plain water looks white. But after you put powdered soap in the water, the light makes a bluish streak across the glass and casts a spot of reddish light on the paper.

You already know that white light is actually a combination of every color (see pages 48-49). Soap particles in the water scatter that light, causing different colors to come out of the glass at different places. The bluish streak of light shows that blue light waves bounce out the front of the glass towards you. Red light waves go through the particles to make the reddish spot on the paper.

What does a glass of soapy water have to do with the sky? The sky also has particles in it: ice crystals, dust, pollen, smoke. When you look at the air between your eyes and this page, it seems very clear. But there are more than 80 km (50 miles) of air above you, with enough particles to scatter some of the Sun's white light. When the Sun is high in the sky, the blue light bounces through the air into your eyes, making the sky appear blue.

At dawn and dusk, sunlight reaches your eyes through even more air because it comes through the atmosphere at an angle. With the colors scattered differently, the Sun looks red orange. Lots of pollution particles in the air also make the Sun look orange.

Add more soap powder to the water. Does the color change? Is the light that gets through as strong?

FROZEN BUBBLE

You can see a bubble freeze if the temperature outside is below 0°C (32°F), and there's not much wind. Grab a cup of Bubble Brew, your Loop (see page 22), and your winter coat. Blow a small bubble and catch it on the Loop. Then carry the bubble outside.

The soapy water in your bubble starts to freeze at one spot, often where a small snowflake sticks to it. The frozen area spreads — it looks like a gray patch swirling on the bubble's surface. Finally, the bubble breaks because the ice is too stiff to stretch, and the frozen section floats off like a big snowflake.

53

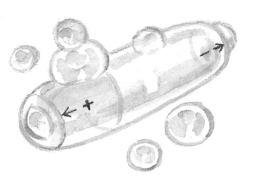

Soapy battery

With soap solution, aluminum, and one cent, you can make a real current event.

You'll need:

- stereo headphones with a jack
- 2 thin wires, each 20 cm (8 inches) long
- a saucer of Bubble Brew (see page 21)
- a shiny copper penny
- aluminum foil

1 Put on the headphones. Hold up the jack that plugs into your stereo. You'll see two thin colored rings dividing the metal into three sections. Wrap the end of one wire tightly around the bottom section.

2 Lay the other end of that wire in the Bubble Brew. Drop the penny on top of it. The penny should be submerged and touching the wire.

3 Wrap the aluminum foil around one end of the second wire. Place the aluminum in the Bubble Brew 0.3 cm (1/8 inch) from the penny.

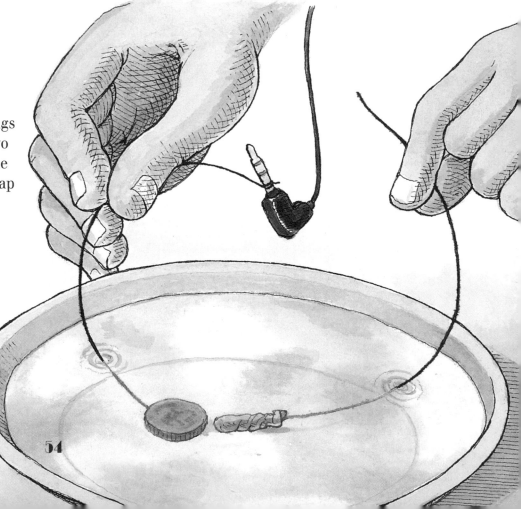

4 Touch the other end of the second wire to the top of the headphone jack, then to the middle section. What do you hear?

WHAT HAPPENS?

As soon as the wire touches the stereo jack, you hear a crackle in one ear. Touching the other section makes your other earphone crackle. If you don't hear anything, make sure that the wires touch separate rings on the jack, that the wires touch the penny and the aluminum foil, and that the penny and foil are close together.

Headphones usually take electricity from your stereo system and convert it into sound for your ears. But these headphones pick up electricity from the soap solution. It's not a big charge, though: the headphones turn it into only a quiet crackle, and it can't hurt you.

Electricity is the movement of the tiny particles called electrons. In this experiment, they move when the top layer of aluminum molecules combines with pieces of the molecules dissolved in the soap solution. Each time they combine, there's an extra electron. These electrons flow up one wire, through your headphones, back down the other wire, and into the copper penny. Then they connect with more pieces of molecules in the liquid, allowing the aluminum to keep reacting and the electrons to keep flowing.

Eventually the entire surface of the aluminum combines with the dissolved molecules, so it can't react any more. But until then, this Bubble Brew contraption is a battery, creating an electrical current.

Clean off the wires, aluminum, and penny. Set up the battery again, but this time in a saucer of fresh water. Can you hear as much through the headphones? What if you use salt water or sugar water?
Make sure the jack is dry before putting it back in the stereo.

FAN OF THE OPERA

Do you sit around the house watching the soap? Probably not, but you may watch the "soaps." These continuing daily dramas were first called "soap operas" by American magazines in 1939, when they were on the radio. Soap companies bought most of the commercials on these shows because they were popular with homemakers, who bought the most soap.

Copy a cartoon

Here's a way to copy a picture from a newspaper, and you don't need any talent for drawing.

You'll need:

- 250 mL (1 cup) water
- 50 mL (1/4 cup) turpentine or mineral spirits
- a scrap of soap the size of one fingernail, or a squirt of liquid soap
- a glass jar with a tight lid
- today's newspaper, after everybody has read it
- scissors
- a paint brush
- 2 sheets of thick white paper
- a metal tablespoon

1 Mix the water and turpentine in the jar. Add the soap. Cap the jar, and shake it hard for three minutes. You need to make sure the ingredients mix together into a cloudy liquid.

2 Clip some pictures that you want to copy from the newspaper. The pictures that copy best have thick lines, bold colors, and no letters or numbers printed on them.

3 Dip your brush in the liquid. Paint the whole picture with it so the paper is wet.

4 Lay the picture face down on one sheet of white paper. Put the other sheet of paper on top of it.

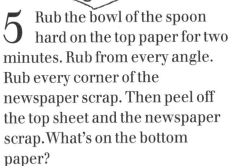

5 Rub the bowl of the spoon hard on the top paper for two minutes. Rub from every angle. Rub every corner of the newspaper scrap. Then peel off the top sheet and the newspaper scrap. What's on the bottom paper?

WHAT HAPPENS?

Some ink comes off the newspaper and sticks to your white paper, giving you a mirror image of the picture you cut out. Newspaper inks stick tight to paper, but the turpentine-and-water mixture dissolves these inks, and pressing on the newspaper with the spoon squeezes them onto the paper below. (Not even this mixture can loosen inks that have been dry for three or four days, so use a freshly printed newspaper.)

This mixture works because water dissolves some of the inks, and turpentine dissolves the rest. A little bit of soap is enough to make these two ingredients combine smoothly so they don't smear the inks or leave any behind. Without soap, water and turpentine don't mix because turpentine is a kind of oil, one that comes from tree resin. (Mineral spirits is another kind of oil, but it comes from the earth.)

How much of the picture can you pick up by painting it with water only? Turpentine only? Water and turpentine but no soap?

Try yesterday's newspaper, and the one from the day before. Try the Sunday comics (if you have a paper route, you know that the comics are printed well before Sunday). How well does the mixture work on drier inks?

57

Bathtub rings

Where do bathtub rings come from? Actually, they come from soap, which can leave behind as much as it washes away.

You'll need:

- 250 mL (1 cup) water
- 2 bowls
- a spoon
- green food coloring
- 3 paper coffee filters or paper towels folded into cones
- a filter holder or a funnel
- a timer
- 30 mL (2 tablespoons) Epsom salts
- liquid soap

1 Put the water in one bowl, and stir in a drop of green food coloring.

2 Put a filter in the filter holder, and pour the water through it into the other bowl. With the timer, time how long it takes the water to flow through. Set the wet filter aside.

3 Add 30 mL (2 tablespoons) of Epsom salts to the water, and stir until the salts dissolve. Does it look different from before?

4 Pour the salty water through another filter into the first bowl. How long does it take for the water to flow through? Set the second wet filter aside.

5 Add three squirts of liquid soap to the water and stir. Let the bowl stand for 15 minutes. Does the water look different from before?

6 Pour the water through the third filter into the second bowl. How long does it take the water to flow through this time?

7 Remove the third wet filter and look at it alongside the other two. Does it look different?

WHAT HAPPENS?

Fresh water flows through a filter quickly, and water with Epsom salts dissolved in it also flows quickly. But water with both Epsom salts and soap takes a long time to flow through. Furthermore, it's cloudy and it leaves a green coating on the paper filter.

Water with a lot of minerals, such as Epsom salts, dissolved in it is called hard water. Many places in North America have naturally hard water, and it's very difficult to remove the minerals because they stick so well to the water molecules. Hard water looks just like water without minerals — until you add soap. The dissolved minerals combine with the dissolved soap to make a new chemical that doesn't dissolve, called a precipitate. You can see the precipitate as flecks in the cloudy water, and as the coating left behind on your filter — or your bathtub.

How can you test that the soap alone isn't responsible for the clogging?

Add borax to hard water. Borax grabs the dissolved minerals. When you add soap to hard water with borax, do you see as much precipitate?

SHAMPOOS

Shampoos are made with detergents instead of soap because in hard water the detergents don't make a precipitate that would stick to people's hair. (Test this out by adding shampoo to hard water and comparing it to hard water and soap.) Detergent isn't the biggest ingredient in most shampoos, though. Read your shampoo's label: the biggest ingredient is probably water.

59

Sudsboat

This toy boat can sail across your bathtub, but it's not powered by wind or oars or motors — it runs on soap!

You'll need:

- a leftover plastic-foam meat tray
- scissors
- liquid soap or detergent
- a bathtub of fresh water

1 Cut two boat shapes, as shown, out of the meat tray.

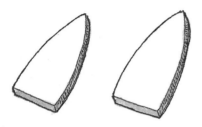

2 Choose one of the boat shapes to be your Sudsboat. Put a dab of soap on the middle of its back edge. Set the boat aside for the soap to dry.

3 Launch the second boat in the tub as your control (see page 7). How fast does it sail?

4 Drop the Sudsboat in the tub next to the control boat. Watch it take off! How does your control boat react?

WHAT HAPPENS?

The Sudsboat moves as soon as you launch it. The soap on its back breaks the bath water's surface tension, causing the top layer of the water to pull back in all directions where the soap touches it. As the water holding up the front of the Sudsboat pulls away, it drags the boat along, too. And as the boat moves through the water, the soap on its back cuts more surface tension, like a zipper unzipping a jacket. After one or two voyages, the Sudsboat won't move any more. There's so much soap in the water that it doesn't have enough surface tension left to unzip. Time to take your bath!

How does the Sudsboat move if, instead of dabbing the soap in the middle of its back edge, you dab the soap on one corner?

What happens if you launch the Sudsboat so that it's pointing the opposite direction from the control boat?

Tips for super science projects

So you're investigating soap for your project. Good idea! Now develop that idea into a terrific project.

Choose a topic

Your first step is to turn your "soapy" idea into a specific topic.

1. Look through *Soap Science* and list any topics that make you curious. Then check over your list and choose the topic that interests you most.
2. Have you seen any newspaper or magazine articles about soap? Are there experiments in *Soap Science* that will help you learn more?
3. Do a project based on one of your hobbies. Do you like soccer but hate grass stains? Can any soap or soap mixtures dissolve these stains?

Choose the type of science project

There are two types of science projects you can choose from:

1. Non-experimental projects
This kind of project doesn't involve an experiment, but still lets you discover the answers to questions you wonder about. You can:
a) show how a scientific principle works (see surface tension, page 10).
b) explain a technology or demonstrate a technique (make soap, page 18).
c) observe and/or collect data (check out bubbles and breezes, page 44).
d) organize a collection (create bubble blowers and frames).

2. Experimental projects
If you like experimenting, this is the type of project for you. Be sure to:
• Investigate questions using the *scientific method* (see below).
• Use a control (page 7) and have a large sample (page 45).

How about investigating:
What Bubble Brew makes the longest-lasting bubbles (page 21)?
How else are soap and detergent different (page 14)?
How can you add color to homemade soap (page 18)?
What frame or air conditions give you the biggest bubbles (page 35)?
How else can you find out about soap's effect on water's surface tension (page 60)?

Scientific method

1. State your *purpose*: say what question you want to answer.
2. Develop a *hypothesis*: make a guess about your experiment's result.
3. Write out your *procedure*: describe what you're going to do.
4. Record your *results*: carefully observe and collect your data.
5. Make *conclusions*: decide what your results mean.

Get organized

Now you're ready to make some plans and lists.

• Write out a step-by-step plan of what you will do in your project.
• Get an okay from an adult that your project is safe.
• List all equipment and materials you need and where you'll get them.
• Create a schedule. Do you have time to do all you have planned?

Get the info

A library is a good place to start researching. You can also try:

Businesses
Conservation areas
Government departments
Hospitals
Museums
Research institutes
Science centers
TV/Radio stations
Zoos

When you ask for information:

Be specific. Ask for material on your particular topic, not just on soap.

Be prepared. Write your questions out ahead of time. Check that they aren't just yes-and-no questions — you'll get more information that way.
Be smart. Call early in case the people you need are away or busy. Take notes as they talk and ask for their names so you can call back if necessary.
Be polite. Write a note afterwards to thank experts for their help.

Ready, set, go!

As you're gathering your material and results:

1. Perform all your experiments carefully and safely.
2. Record your results immediately — make drawings, photos, videos, etc.
3. Be organized — that makes it easier if you have to re-do anything.

Display and present your project

Now put all your material together in a great-looking package.

• Plan your display by sketching it first. How can you make it look better?
• Be as neat as possible.

• Use bright ink and paper. Is the lettering easy to read?
• Check spelling and punctuation.

If you're presenting your project:

• Rehearse with a friend. Ask her for questions, then try to answer them.
• Use notes — but don't just read them. Be sure to make eye contact with your audience.
• Demonstrate part of your project to get your audience really involved.

That's it! Follow these tips and you're on the way to completing your best science project ever.

Glossary

acid: chemical that gives up electrons in a chemical reaction. Acids taste sour.

atom: tiny piece of a molecule. It contains protons and electrons.

bacteria: small, one-celled life forms. Some bacteria can make people sick; some eat people's sweat.

base: chemical that picks up electrons in a chemical reaction. Bases taste bitter.

carbon dioxide: invisible gas that animals (including humans) breathe out. It is also created by mixing vinegar and baking soda.

chemical reaction: combination of different molecules to make new molecules

control: part of an experiment that shows scientists how things normally behave. Scientists compare what the control shows them to what they see when they change something in the experiment.

current: flow of air, water, etc., in a particular direction. Electrons flowing create an electrical current.

detergent: chemical used as a substitute for soap. There are many kinds of detergent, some much stronger than soap.

dissolve: to mix a chemical evenly into a liquid, breaking its molecules into pieces

electron: tiny particle in an atom that has a negative electrical charge. Moving electrons create electricity.

evaporate: to turn from liquid to gas, either by boiling or by floating off into the air one molecule at a time

fat: oily animal (or plant) tissue, usually white or yellow

film: very thin layer of a liquid or solid

lye: water mixed with wood ash. Lye is a strong base.

molecule: smallest particle of a chemical, built from atoms

proton: tiny particle in an atom that has a positive electrical charge

sample: number of times a scientist tries a particular experiment. The bigger the sample, the more certain the scientist can be about the results of the experiment.

soap: chemical created from the reaction of fat and lye. There are many kinds of soap.

solution: liquid with other molecules dissolved evenly in it

sphere: three-dimensional shape that's perfectly round, such as a ball or bubble

surface area: total measure of a shape's outside surfaces

surface tension: force created along a liquid's surface by the connections between the liquid's molecules

Index